Exploring the Elegance of Pure Mathematics: Unveiling the Beauty of Abstract Concepts

Welcome to a journey of discovery through the realm of pure mathematics, a realm where abstract ideas take shape, elegance is paramount, and the pursuit of truth knows no bounds. In "Exploring the Elegance of Pure Mathematics: Unveiling the Beauty of Abstract Concepts," we invite you to delve into a world of thought-provoking abstractions, intricate patterns, and profound insights that transcend time and space.

Pure mathematics is the language of ideas and imagination, a discipline where solutions are born from the interplay of logic, creativity, and intuition. It is a playground for those who relish the pursuit of truth for its own sake, where numbers, symbols, and structures become tools to unravel the mysteries of the universe.

This book is a guide to navigating this captivating landscape, designed for both the curious novice and the seasoned mathematician seeking deeper insights. From the very essence of numbers to the intricacies of abstract algebra, from the geometries of imagined worlds to the harmonic symphonies of number theory, each chapter presents a canvas of concepts waiting to be explored.

Through this exploration, you will encounter the beauty of elegant proofs that unfold like poetry, revealing the symmetry and order hidden within seemingly complex constructs. You will witness the symphony of mathematical objects harmonizing across disciplines, as concepts from algebra, analysis, geometry, and beyond come together in a unified orchestra of ideas.

While pure mathematics may seem distant from the practicalities of our everyday lives, its impact resonates across all fields of human endeavor. It fuels advancements in science, engineering, cryptography, and technology, driving innovation and shaping the way we perceive the world.

Throughout this journey, we will encourage you to embrace the challenges of abstraction, to cultivate your problem-solving skills, and to find inspiration in the profound beauty of mathematical truths. Prepare to marvel at the symmetries that bind nature, to ponder the depths of infinity, and to unravel the mysteries that have perplexed minds for centuries.

Whether you are an aspiring mathematician, a lover of ideas, or simply someone curious about the universe, "Exploring the Elegance of Pure Mathematics" invites you to embark on a voyage of intellectual curiosity and wonder. As you navigate the pages that follow, may you discover not only the elegance of pure mathematics but also the elegance of thought itself.

Introduction: The Enchantment of Pure Mathematics

- Allure and significance of pure mathematics
- Abstract nature and foundational role of pure mathematical concepts

Chapter 1: The Essence of Numbers: Number Theory
- Beauty of prime numbers, divisibility, and arithmetic progressions
- Mysteries of the Riemann Hypothesis and the Goldbach Conjecture

Chapter 2: Beyond the Finite: Set Theory and Infinity

- Concept of infinity and its counterintuitive properties
- Foundations of set theory and the continuum hypothesis

Chapter 3: The Power of Patterns: Group Theory
- Investigating symmetries and transformations through group theory
- Showcasing the Rubik's Cube as an example of group theory application

Chapter 4: The Art of Logic: Mathematical Logic
- World of propositional and predicate logic
- Connections between logic and computational complexity

Chapter 5: The Beauty of Relationships: Functional Analysis
- Functional spaces, operators, and Banach spaces
- Role of functional analysis in modern physics and engineering

Chapter 6: Exploring Infinitesimals: Calculus and Analysis
- Development of calculus and its profound impact on mathematics and science
- Concepts of limits, continuity, derivatives, and integrals

Chapter 7: Dissecting Discreteness: Discrete Mathematics
- Graphs, networks, and discrete structures
- Algorithms, combinatorics, and applications in computer science

Chapter 8: The Art of Arrangements: Combinatorics
- Counting principles, permutations, combinations, and the Pigeonhole Principle
- Combinatorics in cryptography and probability theory

Chapter 9: Finite Wonders: Finite Mathematics
- Finite sets, logic circuits, and game theory

- Applying finite mathematics to decision-making and optimization

Chapter 10: Fractals: Nature's Hidden Patterns
- Beauty of fractals and self-similarity in mathematics and nature
- Mandelbrot Set and fractal geometry's influence on art and design

Conclusion: The Unending Journey of Discovery

- Diverse landscape of pure mathematics
- Continuous nature of mathematical exploration and its impact on understanding the universe

Allure and significance of pure mathematics

The allure and significance of pure mathematics lie in its abstract beauty and its profound impact on shaping our understanding of the world and the universe. Pure mathematics is a realm of thought where ideas are distilled to their most fundamental essence, stripped of practical applications yet brimming with intellectual elegance. Here's a closer look at the allure and significance of pure mathematics:

1. Exploration of Universal Truths: Pure mathematics is a quest for universal truths that transcend time and culture. It seeks to uncover timeless principles and relationships that govern the underlying structure of reality, regardless of practical applications.

2. Intellectual Creativity: Just as artists create masterpieces on canvas, mathematicians create intricate landscapes of thought. The creativity in formulating and proving theorems, inventing new concepts, and exploring uncharted territories is at the heart of pure mathematics.

3. Symmetry and Beauty: The study of pure mathematics often reveals symmetries and patterns that evoke a sense of beauty and wonder. These symmetries can be found in shapes, equations, and the structures that underlie the universe.

4. Intellectual Challenge: The abstract nature of pure mathematics presents intellectual challenges that beckon the curious and the determined. It's a realm where complex problems are tackled through logic, reasoning, and creativity, fostering mental agility and resilience.

5. Foundation for Applied Sciences: While pure mathematics may not always have immediate practical applications, it forms the foundation for applied sciences such as physics, engineering, and computer science. Theorems proven purely for their elegance often find themselves intertwined with real-world advancements.

6. Exploration of Infinity and Beyond: Pure mathematics allows us to explore concepts that defy intuition, such as infinity and other abstract notions. This exploration expands our understanding of the limits of human comprehension and the boundless possibilities of mathematical thought.

7. Influence on Philosophy: Pure mathematics has influenced philosophical debates about the nature of reality, knowledge, and the relationship between the abstract and the concrete. It challenges us to consider the philosophical implications of mathematical truths.

8. A Playground for Intellectual Curiosity: Mathematicians often explore pure mathematics for the sheer joy of intellectual pursuit. It's a playground where curiosity is the driving force, leading to unexpected discoveries and deep insights.

9. Cultural and Historical Significance: Throughout history, pure mathematics has been a hallmark of intellectual achievement. It has transcended cultures, civilizations, and epochs, leaving a legacy of knowledge and wisdom for future generations.

10. Connection to the Universe: Pure mathematics provides a language to describe the laws of the universe. The equations that describe physical phenomena often originate from pure mathematical ideas, revealing a profound connection between abstract concepts and observable reality.

In essence, the allure of pure mathematics lies in its ability to unveil the beauty of abstraction, to engage the intellect in profound ways, and to illuminate the hidden symmetries that shape our universe. Its significance resonates through the pursuit of truth, the advancement of human knowledge, and the eternal quest to understand the fundamental nature of existence.

Abstract nature and foundational role of pure mathematical concepts

The abstract nature and foundational role of pure mathematical concepts are central to the discipline's beauty, power, and influence. Pure mathematics explores ideas that transcend the constraints of the physical world, focusing on the inherent properties of numbers, shapes, structures, and relationships. Here's a deeper look at the abstract nature and foundational role of pure mathematical concepts:

1. Abstraction Beyond Reality: Pure mathematics deals with concepts that are not tied to specific physical objects or phenomena. It ventures into the realm of the abstract, where ideas exist independently of their real-world manifestations.

2. Universal Applicability: The abstract concepts in pure mathematics often have applications that extend far beyond their initial context. They provide a framework for understanding diverse phenomena and making connections across various fields of science and engineering.

3. Formulating Theories and Hypotheses: Mathematicians develop theories and hypotheses based on abstract concepts. These theories serve as frameworks for exploring relationships, making predictions, and generating new insights.

4. Unifying Different Disciplines: Abstract mathematical concepts often serve as a common language between different disciplines. Ideas from pure mathematics, such as group theory, topology, and number theory, find applications in physics, computer science, and even art.

5. Setting the Stage for Applied Mathematics: Pure mathematical concepts lay the foundation for applied mathematics. Ideas developed for their intrinsic elegance often turn out to be incredibly useful in solving real-world problems, driving technological innovation.

6. Discovering Unexpected Connections: The abstract nature of pure mathematics allows mathematicians to discover unexpected connections between seemingly unrelated concepts. These connections can lead to novel insights and breakthroughs.

7. Revealing Hidden Symmetries: Abstract concepts frequently uncover symmetries and patterns that may not be immediately evident in the physical world. These symmetries provide insights into the fundamental order of the universe.

8. Pursuit of Fundamental Truths: Pure mathematics is concerned with seeking fundamental truths about the nature of numbers, geometry, logic, and more. The pursuit of these truths contributes to the development of the broader mathematical landscape.

9. Encouraging Creative Thinking: Working with abstract concepts encourages mathematicians to think creatively and explore unconventional approaches to problem-solving. This fosters a spirit of innovation and inventiveness.

10. Cultivating Intellectual Curiosity: The abstract nature of pure mathematics invites intellectual curiosity. Mathematicians are driven to explore the uncharted territories of abstract spaces, seeking to understand the inherent properties of mathematical objects.

11. Influence on Philosophy and Epistemology: Pure mathematical concepts have influenced philosophical debates about the nature of reality, knowledge, and the relationship between the abstract and the concrete. They challenge our perceptions of the world and the limits of human understanding.

In summary, the abstract nature of pure mathematical concepts allows us to transcend the limitations of the physical world and engage with the fundamental truths that underlie reality. These concepts serve as building blocks for theories, connect different fields of study, and inspire innovation across various disciplines. Through their abstract beauty, they reveal the elegance and order that permeate the universe.

Beauty of prime numbers, divisibility, and arithmetic progressions

The beauty of prime numbers, divisibility, and arithmetic progressions lies in their elegance, simplicity, and profound influence on the world of mathematics and beyond. These concepts are foundational to number theory and have captivated mathematicians for centuries. Let's explore the allure of each:

1. Prime Numbers: Prime numbers are the building blocks of the natural numbers, and their beauty lies in their uniqueness and role in number theory. A prime number is a positive integer greater than 1 that has no positive divisors other than 1 and itself. For example, 2, 3, 5, 7, and 11 are prime numbers.

- **Uniqueness:** Every prime number is distinct, contributing to the uniqueness of the prime factorization of numbers. This uniqueness forms the basis of cryptography and number theory.
- **Distribution:** The distribution of prime numbers is mysterious and irregular. There's no simple formula to predict exactly where prime numbers will appear, which adds to their allure.
- **Goldbach Conjecture:** The Goldbach Conjecture states that every even integer greater than 2 can be expressed as the sum of two prime numbers. This unsolved problem adds an element of intrigue to prime numbers.

2. Divisibility: Divisibility is the concept of one number being evenly divisible by another. The beauty of divisibility lies in its foundational role in arithmetic and its applications in various mathematical areas.

- **Greatest Common Divisor (GCD):** The GCD of two numbers is a fundamental concept that finds applications in simplifying fractions, solving linear Diophantine equations, and more.
- **Euclidean Algorithm:** The Euclidean Algorithm for finding the GCD of two numbers is elegant and efficient, showcasing the beauty of mathematical algorithms.

3. Arithmetic Progressions: An arithmetic progression is a sequence of numbers in which the difference between consecutive terms is constant. These progressions have a simple yet elegant structure that underlies various mathematical and real-world phenomena.

- **Common Difference:** The constant difference between terms in an arithmetic progression gives rise to a predictable pattern, making them easy to comprehend.
- **Sum of Arithmetic Series:** The sum of the first n terms of an arithmetic progression has a simple formula, allowing for quick calculations and insights.
- **Applications:** Arithmetic progressions have applications in a range of fields, from finance (calculating interest) to physics (describing motion with constant acceleration).

The allure of prime numbers, divisibility, and arithmetic progressions lies in their fundamental role in mathematics, their simplicity combined with complexity, and their applications in diverse areas. These concepts reveal the underlying order and patterns in numbers, inviting exploration, discovery, and a deep appreciation for the elegance of mathematical structures.

Mysteries of the Riemann Hypothesis and the Goldbach Conjecture

The Riemann Hypothesis and the Goldbach Conjecture are two of the most captivating unsolved problems in the realm of number theory. They have perplexed mathematicians for centuries, revealing the intricate depths of mathematical mysteries and the challenges that come with their pursuit.

1. The Riemann Hypothesis: The Riemann Hypothesis is a conjecture regarding the distribution of nontrivial zeros of the Riemann zeta function. Proposed by German mathematician Bernhard Riemann in 1859, this hypothesis is often considered one of the most important unsolved problems in mathematics.

Statement of the Hypothesis: The Riemann zeta function, denoted by $\zeta(s)$, is a complex function that holds deep connections to the distribution of prime numbers. The Riemann Hypothesis suggests that all nontrivial zeros of the zeta function lie on a specific line in the complex plane, known as the "critical line." This critical line is the line where the real part of the complex number s is 1/2.

Significance: The Riemann Hypothesis is closely tied to the distribution of prime numbers and has implications for various areas of mathematics, including number theory and mathematical physics. Its resolution would offer insights into the mysterious patterns of prime numbers and could have far-reaching consequences in cryptography and coding theory.

2. The Goldbach Conjecture: The Goldbach Conjecture is

an ancient hypothesis proposed by German mathematician Christian Goldbach in a letter to Euler in 1742. It pertains to the representation of even integers as the sum of two prime numbers.

Statement of the Conjecture: The Goldbach Conjecture posits that every even integer greater than 2 can be expressed as the sum of two prime numbers. In other words, for every even number n > 2, there exist prime numbers p and q such that n = p + q.

Significance: While the Goldbach Conjecture may seem elementary in nature, it remains an open problem despite extensive computational verification for even numbers up to incredibly large values. Solving the conjecture would offer deeper insights into the distribution of prime numbers and the properties of additive number theory.

Both the Riemann Hypothesis and the Goldbach Conjecture showcase the allure of mathematical mysteries that continue to captivate mathematicians and researchers. They serve as reminders of the vastness of mathematical exploration and the enigmatic nature of certain concepts. The pursuit of these unsolved problems demonstrates the resilience, creativity, and determination of mathematicians who seek to unlock the secrets hidden within the fabric of numbers.

Concept of infinity and its counterintuitive properties

The concept of infinity is both fascinating and counterintuitive, challenging our understanding of the finite world we inhabit. It's a notion that has intrigued philosophers, mathematicians, and thinkers for centuries, giving rise to paradoxes, deep philosophical questions, and new mathematical ideas. Let's explore the concept of infinity and its counterintuitive properties:

1. Boundless Nature: Infinity represents an unbounded, limitless, and endless quality that cannot be reached or exhausted. It's a notion that extends beyond any finite quantity, distance, or magnitude.

2. Countability and Uncountability: One of the counterintuitive aspects of infinity is that not all infinities are the same size. Some sets, like the natural numbers (1, 2, 3, ...), are countably infinite, meaning their elements can be listed in an order. However, the real numbers between any two integers are uncountably infinite, defying attempts to establish a one-to-one correspondence with the natural numbers.

3. Hilbert's Paradox of the Grand Hotel: This thought experiment, introduced by mathematician David Hilbert, illustrates the counterintuitive nature of infinity. In a hypothetical hotel with infinitely many rooms, each occupied by a guest, it's possible to accommodate an infinite number of new guests by shifting the occupants to different rooms, even though the hotel is already "full."

4. Zeno's Paradoxes: Ancient Greek philosopher Zeno proposed paradoxes that involve infinity, such as the famous Dichotomy Paradox. In this paradox, a runner attempting to reach a destination is said to cover half the remaining distance in each step, theoretically requiring an infinite number of steps to complete the journey.

5. Infinity in Calculus: Calculus explores limits and the behavior of functions as they approach infinity. The idea of limits enables us to work with infinitely small and infinitely large quantities, facilitating the study of curves, areas, and rates of change.

6. Infinite Series: Infinite series are sums of an infinite number of terms. Some infinite series converge to a finite value, while others diverge to infinity. The study of series leads to the discovery of surprising results, such as the sum of the infinite geometric series $1 + 1/2 + 1/4 + 1/8 + \ldots$ equalling 2.

7. Paradoxes of Infinity: Infinity gives rise to paradoxes that challenge our intuition. Examples include Thomson's Lamp Paradox, which involves flipping a lamp an infinite number of times in a finite period, and the Banach-Tarski Paradox, which suggests that a solid ball can be divided into a finite number of pieces that can be rearranged to form two identical copies of the original ball.

8. Cantor's Work: German mathematician Georg Cantor made significant contributions to our understanding of infinity. He introduced the concept of different sizes of infinity, demonstrated the existence of uncountably infinite sets, and introduced the concept of cardinality to compare the sizes of sets.

In summary, the concept of infinity challenges our intuition and invites us to explore the boundaries of human understanding. It leads to paradoxes, deep philosophical questions, and mathematical insights that continue to shape our perception of the universe. Infinity reminds us that while we may dwell in a finite world, the realm of mathematics allows us to grapple with the infinite and contemplate its enigmatic properties.

Foundations of set theory and the continuum hypothesis

The foundations of set theory provide a framework for understanding the structure of mathematical objects and relationships between them. At the heart of set theory lies the concept of sets, which are collections of distinct elements. Developed by mathematicians such as Georg Cantor and Ernst Zermelo, set theory forms the basis of modern mathematics. The continuum hypothesis is a conjecture within set theory that relates to the cardinality of certain sets. Let's delve into these foundational concepts:

1. Sets and Elements: A set is a fundamental concept in set theory, representing a collection of distinct objects called elements. For example, the set of natural numbers {1, 2, 3, ...} is a collection of individual numbers. Sets can be finite or infinite, and they can contain a variety of elements, such as numbers, letters, or other mathematical objects.

2. Axiomatic Set Theory: Set theory is built upon a set of axioms that define how sets behave and interact. Zermelo-Fraenkel set theory (ZF) is one commonly used axiomatic system. It formalizes concepts like membership, intersection, union, and the creation of new sets through operations.

3. Cardinality and Equivalence: The cardinality of a set represents the "size" of the set in terms of the number of its elements. Two sets have the same cardinality if there exists a one-to-one correspondence (bijection) between them. For example, the sets {1, 2, 3} and {a, b, c} have the same cardinality because there's a bijection between their elements.

4. Cantor's Work: Georg Cantor revolutionized set theory by introducing the concept of different sizes of infinity. He demonstrated that some infinite sets are "larger" than others. For instance, the set of real numbers is uncountably infinite, meaning there's no one-to-one correspondence with the set of natural numbers.

5. The Continuum Hypothesis: The continuum hypothesis is a conjecture formulated by Cantor. It deals with the cardinality of sets related to the real numbers. The hypothesis states that there is no set whose cardinality is strictly between that of the natural numbers and the real numbers.

6. Independence of the Continuum Hypothesis: Kurt Gödel and Paul Cohen later proved that the continuum hypothesis is independent of the standard axioms of set theory. This means that it cannot be proven true or false within those axioms. As a result, the continuum hypothesis is considered undecidable within ZF set theory.

7. Axiom of Choice: The Axiom of Choice is another foundational principle in set theory. It asserts that, given a collection of nonempty sets, it's possible to select exactly one element from each set to form a new set (the "choice set"). The Axiom of Choice has important implications for various areas of mathematics.

8. Set Theory's Impact: Set theory provides a rigorous foundation for mathematical reasoning and serves as a common language for many mathematical disciplines. It plays a crucial role in defining mathematical objects, constructing mathematical proofs, and exploring the logical structure of mathematical concepts.

In summary, the foundations of set theory provide a framework for understanding the relationships between mathematical objects through the concept of sets. The continuum hypothesis, while independent of the standard axioms, showcases the depth and complexity of questions that arise within set theory. As one of the cornerstones of modern mathematics, set theory continues to influence various mathematical disciplines and shape the way mathematicians explore and understand mathematical structures.

Investigating symmetries and transformations through group theory

Investigating symmetries and transformations through group theory uncovers the profound mathematical structure underlying the concept of symmetry in various contexts. Group theory is a branch of abstract algebra that studies the properties and interactions of mathematical sets, called groups, which capture the symmetrical properties of objects and systems. Let's explore how group theory provides a framework for understanding symmetries and transformations:

1. Symmetry and Group Theory: Symmetry refers to the consistent arrangement of parts around a central point or axis. Group theory formalizes the notion of symmetry by representing transformations that preserve the essential properties of an object or system.

2. Definition of a Group: In group theory, a group is defined as a set of elements combined with a binary operation that satisfies four key properties: closure (the product of two elements is also in the group), associativity (the order of operations doesn't matter), identity (there exists an element that acts as an identity), and invertibility (each element has an inverse within the group).

3. Group Actions: A fundamental concept in group theory is the idea of group actions. A group acts on a set when each element of the group corresponds to a transformation of the set's elements. For example, a group of rotations can act on a shape, preserving its symmetry.

4. Symmetry Groups: Symmetry groups describe the collection of all transformations that preserve the symmetrical properties of an object. For instance, the symmetry group of a regular polygon consists of all the rotations and reflections that maintain its shape.

5. Types of Symmetry Groups: Different types of symmetries give rise to various symmetry groups. These include cyclic groups, dihedral groups, permutation groups, and more, each with distinct properties and characteristics.

6. Applications in Geometry: Group theory plays a pivotal role in understanding the symmetrical properties of geometric shapes and structures. It provides a unified framework for classifying and analyzing various patterns of symmetry.

7. Crystallography: In crystallography, group theory is crucial for studying the symmetries of crystals. It helps scientists predict and classify the different possible arrangements of atoms within crystals.

8. Physics and Quantum Mechanics: Group theory is essential in physics, particularly in quantum mechanics. It describes the symmetries of physical systems, which in turn help determine the properties and behavior of particles and interactions.

9. Abstract Algebraic Structures: Group theory is part of a broader context of abstract algebra, which studies algebraic structures beyond the traditional arithmetic operations. It connects symmetries with algebraic properties.

10. Cultural and Artistic Significance: The study of symmetries and group theory extends beyond mathematics. It influences art, design, and aesthetics, as artists and architects draw inspiration from the symmetrical patterns present in nature and mathematics.

In summary, group theory provides a powerful framework

for investigating symmetries and transformations in diverse disciplines. By formalizing the concept of symmetry through mathematical structures, group theory reveals the underlying order and beauty in the symmetrical properties of objects, systems, and the world around us.

Showcasing the Rubik's Cube as an example of group theory application

The Rubik's Cube is a classic and captivating example of how group theory can be applied to solve complex puzzles and understand symmetrical structures. The Cube's movements and transformations form a group that aligns with the principles of group theory. Let's delve into how the Rubik's Cube exemplifies the application of group theory:

1. Symmetries of the Cube: The Rubik's Cube has various symmetrical properties, including rotations along its axes and reflections across its faces. These symmetrical operations are preserved throughout the solving process.

2. Group of Transformations: The set of all possible moves and rotations that can be performed on the Rubik's Cube forms a group. This group represents the collection of transformations that maintain the Cube's shape and arrangement of colors.

3. Group Elements and Operations: Each move on the Rubik's Cube, such as rotating a face 90 degrees clockwise, represents an element of the group. The combination of these moves corresponds to the group's binary operation, where performing one move after another is equivalent to performing a single move.

4. Identity and Inverses: The identity element in the Rubik's Cube group is the state where all faces are in their solved positions. Each move has an inverse move that undoes its effect, maintaining the Cube's symmetry.

5. Group Properties: The Rubik's Cube group exhibits the key properties of a mathematical group. Moves are closed under composition (performing two moves in sequence results in another valid move), associative (the order of moves doesn't matter), and each move has an inverse.

6. Group Theory in Solving: When solving the Rubik's Cube, one applies a sequence of moves to transform the initial state into the solved state. By manipulating the group's elements through a systematic approach, one can navigate the puzzle's symmetrical space to reach a solution.

7. Algorithms and Patterns: Solving the Rubik's Cube involves using algorithms (sequences of moves) to achieve specific configurations. Group theory helps identify algorithms that solve certain patterns efficiently by exploiting the symmetries of the Cube.

8. Generalizations to Higher Dimensions: The concepts of group theory applied to the Rubik's Cube can be extended to higher-dimensional analogs, such as the 4x4x4 or 5x5x5 cubes. These puzzles exhibit even more intricate symmetrical properties.

9. Educational Significance: The Rubik's Cube provides an engaging way to introduce students to group theory concepts. Exploring the group properties of the Cube can offer a hands-on understanding of abstract algebraic concepts.

10. Recreational Mathematics: Beyond its mathematical significance, the Rubik's Cube remains a popular recreational puzzle that challenges and entertains people of all ages.

In summary, the Rubik's Cube serves as a tangible and accessible example of group theory application in solving puzzles with symmetrical properties. By examining the Cube's transformations and operations through the lens of group theory, mathematicians, enthusiasts, and learners gain insights into the elegance of abstract algebra and its practical implications.

World of propositional and predicate logic

The world of propositional and predicate logic forms the foundation of formal reasoning and the study of logical relationships within statements and arguments. These branches of logic provide essential tools for analyzing, constructing, and evaluating logical structures and arguments in various fields, including mathematics, philosophy, computer science, linguistics, and artificial intelligence. Let's delve into the world of propositional and predicate logic:

Propositional Logic:

1. **Propositions:** Propositional logic deals with propositions, which are declarative statements that can be true or false. These propositions are often represented by letters or symbols.
2. **Connectives:** Propositional logic introduces connectives that allow the construction of compound propositions from simpler ones. Common connectives include "and" (conjunction), "or" (disjunction), "not" (negation), "if...then..." (implication), and "if and only if" (biconditional).
3. **Truth Tables:** Truth tables are used to systematically list all possible truth values of compound propositions based on the truth values of their component propositions and connectives.
4. **Logical Equivalences:** Propositional logic involves the study of logical equivalences, which are statements that have the same truth value under all circumstances. Equivalences facilitate simplification and manipulation of logical expressions.

Predicate Logic:

1. **Predicates and Quantifiers:** Predicate logic extends propositional logic by introducing predicates (functions that return true or false for specific inputs) and quantifiers ("for all" and "there exists") that allow for the expression of statements about variables and their relationships.
2. **Variables and Constants:** Predicate logic introduces variables to represent unspecified objects and constants to represent specific objects. Predicates and quantifiers help define relationships between these objects.
3. **Universal and Existential Quantifiers:** The universal quantifier ($\forall$) asserts that a statement holds true for all elements in a given domain, while the existential quantifier ($\exists$) asserts the existence of at least one element satisfying a condition.
4. **Relations and Functions:** Predicate logic allows for the formal representation of relations and functions, enabling precise descriptions of mathematical structures and properties.
5. **Inference Rules:** Predicate logic involves inference rules that facilitate logical deduction. Modus ponens and modus tollens are examples of common inference rules.

Applications:

1. **Mathematics:** Propositional and predicate logic are essential tools in mathematical proof, allowing mathematicians to rigorously establish the validity of statements and theorems.
2. **Philosophy:** Logic plays a pivotal role in philosophical reasoning, helping to analyze arguments and propositions and determine their validity.
3. **Computer Science:** Propositional logic serves as the foundation for Boolean algebra and digital circuit

design. Predicate logic is essential for programming, formal verification, and database management.

4. **Linguistics:** Logical frameworks are used in analyzing the structure and semantics of natural languages.

5. **Artificial Intelligence:** Propositional and predicate logic are used to represent knowledge and reason in AI systems, enabling machines to make informed decisions and draw logical conclusions.

In summary, propositional and predicate logic provide the tools for formal reasoning and analysis of logical relationships within statements and arguments. Their applications extend to a wide range of disciplines, offering a structured and systematic approach to reasoning, problem-solving, and knowledge representation.

Connections between logic and computational complexity

The connections between logic and computational complexity run deep and offer insights into the limits and possibilities of solving computational problems. Logic provides a formal language for expressing and analyzing the structure of arguments and statements, while computational complexity theory focuses on understanding the efficiency of algorithms in solving computational problems. Let's explore how these two fields are interconnected:

1. Formal Representation of Problems: Logic provides a way to formally represent problems and their solutions. Logical statements can describe the relationships and constraints within a problem, offering a clear and structured way to express the problem's properties.

2. Complexity Classes and Logic: Computational complexity theory defines complexity classes such as P (problems solvable in polynomial time) and NP (problems verifiable in polynomial time). Logic plays a role in characterizing these classes. For example, problems in P can often be expressed using first-order logic, while problems in NP can be verified using polynomial-time verifiers expressed in second-order logic.

3. Expressive Power and Complexity: The expressive power of logical languages can influence the computational complexity of problems. More expressive logic languages can describe more intricate relationships, but they can also lead to problems that are harder to solve (higher complexity).

4. Descriptive Complexity: Descriptive complexity theory explores the relationship between logical expressiveness and complexity classes. It connects specific logical languages to complexity classes, providing a way to analyze the computational complexity of problems based on their logical representation.

5. Model Theory and Complexity: Model theory, a branch of mathematical logic, studies the relationships between mathematical structures and the logical languages that describe them. This connection between model theory and complexity theory helps explain the inherent difficulty of certain problems.

6. Cook-Levin Theorem: The Cook-Levin Theorem, establishing the NP-completeness of the Boolean satisfiability problem (SAT), plays a central role in computational complexity theory. It shows that a problem in NP can be reduced to SAT, which makes SAT a canonical NP-complete problem. This theorem demonstrates the interplay between logic, complexity, and reduction.

7. Logic as a Tool for Analysis: Logical frameworks are used to analyze the complexity of algorithms and problems. By formalizing the structure of problems and the steps of algorithms, researchers can assess the efficiency of algorithms in terms of time and resources.

8. Theoretical Foundations for Algorithms: Understanding the complexity of problems helps guide the development of efficient algorithms. Logic provides a foundation for studying the inherent difficulty of problems and aids in designing algorithms that exploit problem-specific structures.

9. Limitations and Intractability: Logic can reveal the inherent intractability of certain problems. Problems that are proven to be NP-complete have no known polynomial-time solutions, unless P equals NP, a major unsolved question in computer science.

In summary, the connections between logic and computational complexity showcase how the structure of problems, logical languages, and algorithmic efficiency are intertwined. Logic provides a formal way to express problems, while complexity theory explores the limits and complexities of solving those problems algorithmically. This intersection is crucial for understanding the inherent difficulty of computational tasks and for guiding the development of efficient algorithms.

Functional spaces, operators, and Banach spaces

Functional spaces, operators, and Banach spaces are fundamental concepts in functional analysis, a branch of mathematics that studies spaces of functions and their properties. These concepts play a crucial role in understanding the structure, behavior, and convergence of functions, particularly in infinite-dimensional spaces. Let's explore these concepts:

1. Functional Spaces: Functional spaces are sets of functions that share certain properties, making them amenable to analysis. These spaces provide a framework for studying the properties of functions, such as continuity, differentiability, and integrability. Examples of functional spaces include:

- **Lp Spaces:** These spaces consist of functions for which the pth power of the absolute value is Lebesgue integrable over a given domain. L2 is particularly important in contexts involving inner product spaces and Hilbert spaces.
- **Ck Spaces:** These spaces consist of functions that are k times continuously differentiable, with continuous kth derivatives.
- **Sobolev Spaces:** These spaces encompass functions and their derivatives, capturing regularity properties in various contexts, including partial differential equations.

2. Operators: Operators are mappings that transform functions from one space to another. In functional analysis, operators are used to analyze how functions interact with one another, often by mapping them to different functional spaces. Some types of operators include:

- **Linear Operators:** These operators preserve linearity; that is, they satisfy properties like additivity and scalar multiplication.
- **Differential Operators:** These operators involve derivatives and play a significant role in differential equations and calculus of variations.
- **Integral Operators:** These operators involve integrals and are often used in integral equations.
- **Compact Operators:** These are operators that map bounded sets to relatively compact sets.

3. Banach Spaces: A Banach space is a normed vector space that is complete with respect to its norm. Completeness means that every Cauchy sequence in the space converges to a limit within the space. Examples of Banach spaces include Lp spaces, spaces of continuous functions, and spaces of integrable functions. The concept of a Banach space provides a way to extend the notions of convergence and limits to infinite-dimensional spaces.

4. Linear Functionals and Dual Spaces: Linear functionals are linear mappings from a vector space to the real or complex numbers. The dual space of a vector space consists of all linear functionals on that space. Dual spaces have applications in various mathematical and physical contexts.

5. Norms and Inner Products: Many functional spaces come equipped with norms or inner products that quantify the size and distance of functions. Norms and inner products provide a way to measure the behavior and convergence of functions.

6. Applications: Functional analysis has applications in a wide range of mathematical areas, including functional equations, differential equations, harmonic analysis, quantum mechanics, signal processing, and optimization.

In summary, functional spaces, operators, and Banach spaces form the foundation of functional analysis, enabling the study of functions' properties, convergence, and interactions in infinite-dimensional settings. These concepts play a vital role in various mathematical and scientific disciplines, offering a powerful framework for understanding and solving problems involving functions and transformations.

Role of functional analysis in modern physics and engineering

Functional analysis plays a crucial role in modern physics and engineering by providing a powerful mathematical framework for understanding complex systems, modeling physical phenomena, and solving real-world problems. Its applications span a wide range of fields, from quantum mechanics and electromagnetism to signal processing and control theory. Here's how functional analysis impacts modern physics and engineering:

1. Quantum Mechanics: Functional analysis is central to the mathematical formalism of quantum mechanics. Quantum states are represented as vectors in a Hilbert space, a type of functional space equipped with an inner product. Operators, such as Hamiltonians and observables, are represented as linear transformations on these spaces. Spectral theory, a branch of functional analysis, helps analyze the spectrum of observables, energy levels, and quantum states.

2. Quantum Field Theory: Functional analysis plays a significant role in quantum field theory, a framework that extends quantum mechanics to systems with an infinite number of degrees of freedom. The creation and annihilation operators used in quantizing fields are represented using operator algebras and functional spaces.

3. Electromagnetism and Quantum Electrodynamics (QED): In QED, functional analysis helps describe the interactions between electromagnetic fields and charged particles. The concept of the

Fock space, a type of Hilbert space, is used to model states of varying numbers of photons.

4. Partial Differential Equations (PDEs): Many physical phenomena are described by PDEs. Functional analysis techniques, such as the theory of distributions and Sobolev spaces, provide a rigorous foundation for solving and analyzing PDEs arising in fluid dynamics, heat transfer, and wave propagation.

5. Control Theory and Engineering: Functional analysis is crucial in control theory, where it's used to analyze and design control systems for complex dynamical systems. Operator theory helps model and analyze systems with infinite-dimensional state spaces, such as distributed parameter systems and systems described by PDEs.

6. Signal Processing: In signal processing, functional analysis is used to analyze and process signals in various domains (time, frequency, etc.). Transform methods like the Fourier transform and the wavelet transform are defined in terms of functional spaces, allowing signals to be represented and manipulated more effectively.

7. Optimization and Numerical Analysis: Functional analysis plays a role in optimization problems, helping to develop algorithms for finding optimal solutions. It provides tools for analyzing convergence properties and characterizing feasible regions in optimization processes.

8. Materials Science and Engineering: In material science, functional analysis contributes to the modeling of material properties, such as thermal conductivity and mechanical behavior. It's used in analyzing transport phenomena, phase transitions, and properties of complex materials.

9. Computational Fluid Dynamics: Functional analysis helps analyze and solve fluid dynamics problems. The finite element method and finite volume method, widely used in computational fluid dynamics, are based on functional analysis principles.

10. Image Processing and Computer Vision: Functional analysis techniques are used in image processing and computer vision applications, aiding in tasks such as image denoising, restoration, and feature extraction.

In summary, functional analysis provides a unifying framework for tackling the mathematical challenges posed by complex physical and engineering systems. Its concepts and techniques are foundational for modeling, analyzing, and solving problems in modern physics and engineering, enabling scientists and engineers to understand the behavior of complex systems, design innovative solutions, and make advancements in various scientific and technological domains.

Development of calculus and its profound impact on mathematics and science

The development of calculus marks a pivotal moment in the history of mathematics and science. It revolutionized the way we understand and describe change, motion, and complex phenomena. The contributions of Isaac Newton and Gottfried Wilhelm Leibniz in the 17th century laid the foundation for calculus, profoundly impacting both mathematics and various scientific disciplines. Here's how calculus developed and its profound impact:

1. Origins of Calculus: Calculus emerged independently in the work of Newton and Leibniz, who developed similar ideas and notations. They introduced the fundamental concepts of derivatives (instantaneous rates of change) and integrals (accumulated quantities), providing a powerful framework for analyzing continuous change.

2. Differential Calculus: Differential calculus deals with rates of change and slopes of curves. It introduced the concept of the derivative, which measures the rate at which a quantity changes with respect to another. The derivative enables the analysis of motion, optimization, and instantaneous behavior.

3. Integral Calculus: Integral calculus focuses on the accumulation of quantities over intervals. The integral represents the area under a curve, which has applications in computing areas, volumes, and accumulated quantities like work and distance.

4. Fundamental Theorem of Calculus: The Fundamental Theorem of Calculus establishes a deep connection between differentiation and integration. It states that differentiation and integration are inverse operations, allowing the calculation of integrals through antiderivatives and vice versa.

5. Applications in Physics: Calculus revolutionized physics, enabling the precise description of motion, forces, and dynamics. It provided tools for understanding planetary orbits, falling bodies, and fundamental laws like Newton's laws of motion and gravitation.

6. Mathematics Reforms: Calculus led to a transformation in mathematics. It introduced new techniques for solving equations, finding extrema, and investigating properties of curves and surfaces. The rigorous development of calculus laid the groundwork for the growth of analysis and modern mathematical analysis.

7. Calculus of Variations: The calculus of variations, an extension of calculus, addresses optimization problems involving functionals (functions of functions). It has applications in fields such as mechanics, economics, and engineering.

8. Impact on Engineering and Technology: Calculus is essential in engineering and technology, enabling the design and analysis of structures, electrical circuits, fluid dynamics, and more. It's fundamental to disciplines like civil engineering, mechanical engineering, and aerospace engineering.

9. Probability and Statistics: Calculus plays a role in probability theory and statistics. Probability distributions and statistical analysis involve integrals and derivatives, influencing fields such as finance, epidemiology, and data science.

10. Modern Science and Technology: Calculus underpins many modern scientific and technological advancements. It's used in fields ranging from genetics and climate modeling to computer graphics and artificial intelligence.

In summary, the development of calculus sparked a mathematical and scientific revolution, reshaping our understanding of the natural world and providing tools for modeling and analysis across various disciplines. Calculus remains an essential pillar of mathematics, enabling advances in both theory and practical applications that have shaped the course of human knowledge and progress.

Concepts of limits, continuity, derivatives, and integrals

The concepts of limits, continuity, derivatives, and integrals are fundamental building blocks of calculus, forming the backbone of mathematical analysis and its applications in various fields. These concepts provide the tools to understand and describe change, motion, and accumulation in both continuous and discrete contexts. Let's explore each concept:

1. Limits: A limit represents the behavior of a function as its input approaches a particular value. It's a foundational concept in calculus and captures the idea of approaching a value as closely as possible without actually reaching it. Limits are used to define continuity, derivatives, and integrals.

2. Continuity: A function is continuous at a point if its graph has no breaks, jumps, or gaps at that point. In other words, the function's values change smoothly as the input changes. A function is continuous on an interval if it's continuous at every point within that interval.

3. Derivatives: The derivative of a function represents its rate of change at a specific point. Geometrically, it corresponds to the slope of the tangent line to the graph of the function at that point. Derivatives have applications in describing motion, optimization, and the behavior of functions.

4. Integrals: Integrals are used to calculate the accumulated total of a quantity over an interval or region. Geometrically, the integral represents the area under a curve. Integrals have applications in computing areas, volumes, accumulated quantities, and solving differential equations.

5. Fundamental Theorem of Calculus: The Fundamental Theorem of Calculus establishes a deep connection between differentiation and integration. It states that differentiation and integration are inverse operations, allowing the calculation of integrals through antiderivatives and vice versa.

6. Applications:

- Physics: Derivatives and integrals model motion, forces, and dynamics in physics. They're fundamental to understanding concepts like velocity, acceleration, work, and energy.
- Engineering: Calculus is essential in engineering, enabling the design and analysis of structures, electrical circuits, fluid dynamics, and more.
- Economics: Calculus is used in economics to model relationships between variables, optimize resource allocation, and analyze economic behavior.
- Biology: Calculus is used to model growth rates, population dynamics, and biological processes.
- Computer Science: Calculus concepts are applied in computer graphics, algorithms, and simulations.

7. Notation:

- Limits are often denoted using the symbol "lim" followed by the variable approaching a value.
- The derivative of a function $f(x)$ is denoted as $f'(x)$ or dy/dx.
- The integral of a function $f(x)$ is denoted as $\int f(x)\, dx$.

8. Rigorous Definitions: The concepts of limits, continuity, derivatives, and integrals are rigorously defined using mathematical definitions involving epsilon-delta notation for limits and the definition of the derivative as a limit of difference quotients.

In summary, the concepts of limits, continuity, derivatives, and integrals provide a powerful framework for understanding change, accumulation, and the behavior of functions. These concepts are not only fundamental in calculus but also play a significant role in describing and modeling phenomena across a wide range of disciplines.

Graphs, networks, and discrete structures

Graphs, networks, and discrete structures are fundamental concepts in mathematics and computer science that model relationships, connections, and structures in various systems. These concepts provide a versatile framework for representing real-world scenarios, solving problems, and analyzing relationships between entities. Let's explore each of these concepts:

1. Graphs: A graph is a collection of nodes (vertices) and edges that connect pairs of nodes. Graphs are used to model relationships between objects, entities, or concepts. They come in various types:

- **Undirected Graphs:** Edges have no direction, and connections between nodes are symmetric.
- **Directed Graphs (Digraphs):** Edges have a direction, indicating a one-way relationship.
- **Weighted Graphs:** Edges have weights that represent costs, distances, or other measures.
- **Cyclic and Acyclic Graphs:** Cyclic graphs contain cycles (loops), while acyclic graphs have no cycles.
- **Bipartite Graphs:** Nodes can be divided into two disjoint sets, with edges connecting nodes from different sets.

2. Networks: Networks refer to structures where nodes represent entities (such as computers, people, or cities), and edges represent interactions or connections between these entities. Networks have applications in various fields:

- **Social Networks:** Represent connections between individuals in social interactions.

- **Transportation Networks:** Model connections between cities, airports, roads, etc.
- **Communication Networks:** Model connections between devices, like the internet.
- **Biological Networks:** Model interactions between proteins, genes, and other biological entities.

3. Discrete Structures: Discrete structures refer to mathematical objects that are countable and separated, as opposed to continuous structures. They have applications in computer science, cryptography, and other fields:

- **Sets:** Collections of distinct objects, forming a fundamental building block of many mathematical structures.
- **Functions:** Relationships that assign each input in a set to a unique output in another set.
- **Relations:** Generalizations of functions, where pairs of inputs and outputs are related.
- **Combinatorics:** The study of counting and arranging objects in various ways.
- **Number Theory:** The study of properties of integers and their relationships.

4. Applications:

- **Computer Science:** Graphs are used in data structures, algorithms, and network design. Discrete structures underpin programming and cryptography.
- **Transportation and Logistics:** Networks model transportation systems, optimizing routes, and distribution.
- **Social Sciences:** Graphs model social interactions, influence, and information flow.
- **Biology and Medicine:** Networks describe protein interactions, neural connections, and disease spread.
- **Engineering:** Graphs and networks are used in

circuit design, telecommunications, and optimization problems.

5. Graph Theory: Graph theory is a branch of mathematics that studies graphs and their properties. It explores concepts like connectivity, shortest paths, graph coloring, and matching.

6. Discrete Mathematics: Discrete mathematics focuses on countable structures and includes graph theory, combinatorics, number theory, and more.

In summary, graphs, networks, and discrete structures are fundamental mathematical concepts that provide tools for modeling relationships, connections, and structures in diverse scenarios. They play a vital role in computer science, engineering, social sciences, and many other fields, offering insights into the underlying structures of complex systems and enabling problem-solving and analysis.

Algorithms, combinatorics, and applications in computer science

Algorithms, combinatorics, and their applications are essential components of computer science, providing the tools to solve problems efficiently, analyze structures, and design systems that power modern technology. Let's explore each of these concepts and their role in computer science:

1. Algorithms: Algorithms are step-by-step instructions or procedures for solving problems or performing tasks. They are the heart of computer science, enabling computers to process data and execute tasks. Key aspects of algorithms include:

- **Efficiency:** Algorithms strive to be efficient in terms of time and space complexity.
- **Correctness:** Algorithms should produce the correct output for all possible inputs.
- **Design Techniques:** Algorithms can be designed using techniques like divide and conquer, dynamic programming, and greedy algorithms.

2. Combinatorics: Combinatorics is the study of counting, arranging, and selecting objects. It's crucial in various areas of computer science:

- **Permutations and Combinations:** Combinatorial techniques help count possibilities and analyze arrangements.
- **Graph Theory:** Graphs and networks involve combinatorial structures like paths, cycles, and

coloring.

- **Cryptography:** Combinatorial analysis is used to assess the security of cryptographic algorithms.

3. Applications in Computer Science:

- **Data Structures:** Algorithms are used to manipulate and process data stored in structures like arrays, linked lists, trees, and graphs.
- **Sorting and Searching:** Efficient algorithms for sorting and searching are fundamental in databases, information retrieval, and data analysis.
- **Graph Algorithms:** Algorithms for finding shortest paths, network flows, and matching have applications in network design, transportation systems, and social networks.
- **Dynamic Programming:** This technique is applied in optimization problems, such as resource allocation and scheduling.
- **Cryptographic Algorithms:** Encryption, digital signatures, and secure key exchange rely on algorithms from number theory and combinatorics.
- **Artificial Intelligence:** Algorithms drive machine learning, pattern recognition, and decision-making systems.
- **Networks and Routing:** Algorithms determine how data is routed through computer networks and the internet.
- **Database Management:** Algorithms handle efficient storage, retrieval, and manipulation of data in databases.
- **Computer Graphics:** Algorithms render images, simulate physics, and create visual effects in video games and movies.

4. Computational Complexity: Computational complexity theory studies the efficiency of algorithms and the inherent difficulty of computational problems. It classifies problems into complexity classes like P (problems solvable in polynomial time) and NP (problems verifiable in polynomial time).

5. Coding and Encryption: Algorithms play a role in coding theory, error correction, and encryption methods. They enable secure communication and data storage.

6. Optimization: Algorithms find optimal solutions in various contexts, such as resource allocation, network design, and scheduling.

In summary, algorithms, combinatorics, and their applications form the core of computer science. They empower us to solve problems efficiently, analyze complex structures, and create technologies that drive innovation in various domains, from artificial intelligence and cryptography to networking and data analysis.

Counting principles, permutations, combinations, and the Pigeonhole Principle

Counting principles, permutations, combinations, and the Pigeonhole Principle are fundamental concepts in combinatorics that deal with counting and arranging objects in various ways. These concepts play a crucial role in solving problems related to arrangements, selections, and probability. Let's explore each of these concepts:

1. Counting Principles: Counting principles provide systematic approaches for counting the number of outcomes in different scenarios. Two fundamental counting principles are:

- **Multiplication Principle:** If a task can be broken down into multiple independent subtasks, the total number of outcomes is the product of the possibilities for each subtask.
- **Addition Principle:** If a task can be divided into mutually exclusive cases, the total number of outcomes is the sum of the outcomes in each case.

2. Permutations: Permutations refer to the arrangement of objects in a specific order. A permutation of n objects taken r at a time is denoted as P(n, r), and it represents the number of ways to arrange r objects out of n when order matters. Permutations are calculated using formulas like n! / (n - r)!, where n! represents the factorial of n.

3. Combinations: Combinations refer to the selection of objects without considering their order. A combination of n objects taken r at a time is denoted as C(n, r), and it represents the number of ways to select r objects out of n when order doesn't matter. Combinations are calculated using formulas like n! / (r! * (n - r)!).

4. Pigeonhole Principle: The Pigeonhole Principle, also known as the Dirichlet Principle, states that if you distribute more objects (pigeons) into a smaller number of containers (pigeonholes), at least one container must contain more than one object. In other words, if you have more items than available options, some option must occur more than once.

5. Applications:

- **Probability and Statistics:** Permutations and combinations are used to calculate probabilities in various scenarios, such as drawing cards from a deck or selecting a committee.
- **Combinatorial Problems:** Counting principles, permutations, and combinations help solve problems involving arrangements, distributions, and selections.
- **Cryptography:** Combinatorial analysis is used to assess the security of encryption algorithms and cryptographic protocols.
- **Design and Arrangements:** Permutations and combinations have applications in arranging objects, designing codes, and organizing events.

6. Binomial Theorem: The Binomial Theorem is used to expand expressions of the form $(a + b)^n$, where n is a positive integer. It's based on the coefficients of the terms in the expansion, which are determined by combinations.

7. Generating Functions: Generating functions are used to encode sequences of numbers into algebraic expressions. They provide a powerful tool for solving combinatorial problems, particularly those involving sums and sequences.

In summary, counting principles, permutations, combinations, and the Pigeonhole Principle are essential tools in combinatorics, enabling us to calculate arrangements, selections, and probabilities in diverse scenarios. These concepts play a significant role in solving problems across mathematics, probability theory, cryptography, and various areas of science and engineering.

Combinatorics in cryptography and probability theory

Combinatorics plays a crucial role in both cryptography and probability theory, providing essential tools for analyzing security, calculating probabilities, and designing cryptographic algorithms. Let's delve into how combinatorics is applied in these two fields:

Combinatorics in Cryptography:

1. **Key Space Size and Security:** In cryptography, the size of the key space (the number of possible keys) is a critical factor in determining the security of a cryptographic algorithm. Combinatorial analysis helps calculate the number of possible keys and assess the feasibility of brute-force attacks.

2. **Permutations and Substitution Ciphers:** Substitution ciphers involve replacing letters with other letters or symbols. Combinatorial analysis is used to calculate the number of possible permutations in these ciphers and evaluate their security against frequency analysis attacks.

3. **Combinatorial Analysis of Attacks:** Cryptanalysts use combinatorial methods to analyze potential attacks on cryptographic systems. This includes analyzing the complexity of different attack strategies and estimating the time required to break the security of a system.

4. **Block Cipher Design:** Block ciphers involve dividing plaintext into fixed-size blocks and applying

cryptographic transformations. Combinatorial analysis helps design S-boxes (substitution boxes) and analyze their avalanche effect, ensuring that small changes in input lead to significant changes in output.

Combinatorics in Probability Theory:

1. **Counting Outcomes:** Probability theory involves counting the number of favorable outcomes and total possible outcomes. Combinatorial techniques such as permutations and combinations are used to calculate these counts.
2. **Probability Distributions:** Combinatorics is used to define and analyze discrete probability distributions, where each outcome has a certain probability. Binomial and multinomial distributions involve counting the number of ways events can occur.
3. **Combinatorial Probability Problems:** Probability theory addresses problems like coin flips, dice rolls, card draws, and more. Combinatorial analysis helps calculate the probability of specific outcomes and events.
4. **Random Walks and Path Counting:** Random walks involve a sequence of steps in various directions. Combinatorics helps calculate the number of possible paths and analyze the behavior of random walks.
5. **Permutations and Random Arrangements:** Probability problems related to arranging objects or sequences often involve permutations. Combinatorial analysis aids in calculating the probability of specific arrangements.
6. **Birthday Paradox:** The birthday paradox highlights the surprising likelihood of shared birthdays in a group. Combinatorial analysis is used to calculate the probability of this occurrence.

In summary, combinatorics serves as a powerful tool in both cryptography and probability theory. In cryptography, it contributes to the analysis of security, the design of cryptographic systems, and the assessment of attacks. In probability theory, combinatorial techniques help calculate probabilities, analyze distributions, and solve a wide range of probability-related problems. Both fields benefit from the systematic methods that combinatorics offers for counting, arranging, and analyzing outcomes and events.

Finite sets, logic circuits, and game theory

Finite sets, logic circuits, and game theory are distinct mathematical concepts that find applications in various fields, from computer science and engineering to economics and social sciences. Let's explore each of these concepts and their applications:

1. Finite Sets: A finite set is a collection of distinct elements with a countable number of members. Finite sets play a fundamental role in discrete mathematics, computer science, and combinatorics.

Applications:

- **Counting and Combinatorics:** Finite sets are used to model and count objects, arrangements, and combinations in various scenarios.
- **Discrete Structures:** Sets are a foundational concept in discrete mathematics, serving as the basis for relations, functions, and graphs.
- **Database Systems:** Finite sets are used to represent and manipulate data in databases and information systems.

2. Logic Circuits: Logic circuits are systems that perform logical operations on input signals to produce output signals. They form the basis of digital systems, including computers, calculators, and communication devices.

Applications:

- **Computer Architecture:** Logic circuits are used to design the central processing unit (CPU) and memory components of computers.

- **Digital Electronics:** Logic gates, which implement Boolean functions, are used to build complex digital systems.
- **Embedded Systems:** Logic circuits are present in a wide range of devices, from household appliances to industrial machinery.
- **Cryptography:** Boolean logic is used to design secure cryptographic algorithms and protocols.

3. Game Theory: Game theory studies interactions and strategic decision-making in competitive and cooperative situations. It has applications in economics, political science, biology, and more.

Applications:

- **Economics:** Game theory models economic behaviors, such as market competition, pricing strategies, and bargaining situations.
- **Political Science:** Game theory helps analyze political elections, international relations, and policy decisions.
- **Biology:** Evolutionary game theory models behaviors in natural selection and animal interactions.
- **Computer Science:** Game theory is used in algorithmic design, mechanism design, and network protocols.
- **Social Sciences:** Game theory provides insights into decision-making, cooperation, and conflict resolution.

4. Logic Circuits and Finite Sets: Logic circuits often involve representing logical states using finite sets, where each element corresponds to a state (e.g., 0 or 1). Logic gates perform operations on these states, implementing Boolean functions.

5. Game Theory and Finite Sets: In game theory, strategies and outcomes can be represented using finite sets. Players' choices and potential outcomes are often defined by finite sets of options.

In summary, finite sets, logic circuits, and game theory are distinct mathematical concepts with diverse applications. Finite sets provide a foundation for counting and discrete structures, logic circuits underlie digital systems, and game theory analyzes strategic decision-making in various contexts. These concepts demonstrate the versatility of mathematics in addressing complex problems across different fields.

Applying finite mathematics to decision-making and optimization

Finite mathematics plays a significant role in decision-making and optimization problems across various fields. It provides tools and techniques to model, analyze, and solve problems that involve limited resources, constraints, and objectives. Here's how finite mathematics is applied to decision-making and optimization:

1. Linear Programming: Linear programming is a mathematical technique used to find the best possible outcome in a scenario with linear constraints. It's widely applied in various industries for resource allocation, production planning, and supply chain optimization.

2. Network Optimization: Network optimization involves finding the most efficient way to allocate resources in a network, whether it's transportation routes, communication networks, or project scheduling. Techniques like the shortest path algorithm and the maximum flow-minimum cut theorem are used.

3. Allocation Problems: Finite mathematics helps in solving allocation problems, such as distributing resources, assigning tasks, and allocating budgets to maximize overall efficiency and minimize costs.

4. Inventory Management: Finite mathematics is used to optimize inventory levels by considering factors like demand, storage costs, and ordering costs. Techniques like the Economic Order Quantity (EOQ) and the Reorder Point (ROP) help find optimal inventory policies.

5. Facility Location and Siting: When deciding where to locate facilities like factories, warehouses, or service centers, finite mathematics can be used to consider factors such as transportation costs, demand distribution, and access to resources.

6. Project Scheduling: In project management, finite mathematics aids in scheduling tasks, assigning resources, and minimizing project completion time while considering constraints and dependencies.

7. Investment and Portfolio Optimization: In finance, finite mathematics is used to optimize investment portfolios by considering risk, return, and diversification. Techniques like the Markowitz Portfolio Theory help investors make informed decisions.

8. Game Theory Applications: Finite mathematics is applied in game theory to analyze strategic interactions and make decisions in competitive scenarios, ranging from economics to international relations.

9. Resource Allocation in Healthcare: Finite mathematics helps in optimizing the allocation of healthcare resources, such as hospital beds, medical equipment, and staff, to maximize patient care under budget constraints.

10. Transportation and Route Optimization: Finite mathematics is used to optimize transportation routes, vehicle scheduling, and distribution to minimize costs, delivery times, and fuel consumption.

11. Production and Manufacturing: Finite mathematics assists in production scheduling, batch sizing, and quality control in manufacturing environments to optimize resource utilization and reduce waste.

12. Energy Management: In energy production and distribution, finite mathematics helps optimize power generation, transmission, and distribution to meet demand efficiently and minimize costs.

In all these applications, finite mathematics provides a structured framework to make informed decisions while considering constraints, objectives, and available resources. It enables organizations to optimize their operations, allocate resources wisely, and achieve efficient outcomes across various domains.

Beauty of fractals and self-similarity in mathematics and nature

Fractals and self-similarity are captivating concepts that bridge the gap between mathematics and the natural world. These intricate patterns are found throughout nature and mathematics, reflecting a deep interplay between complexity and simplicity. Let's explore the beauty and significance of fractals and self-similarity:

1. Fractals: Fractals are complex geometric shapes that exhibit self-similarity at different scales. This means that as you zoom in on a fractal, you'll encounter similar patterns repeating over and over. Fractals often have fractional dimensions, which adds to their uniqueness.

2. Self-Similarity: Self-similarity refers to the property where a part of an object resembles the whole or other parts of the same object. This property allows fractals to be constructed through recursive processes.

Beauty of Fractals and Self-Similarity:

1. **Intricate Detail:** Fractals reveal astonishing detail and complexity regardless of the scale at which they're observed. This characteristic makes them visually captivating and mathematically intriguing.
2. **Mathematical Elegance:** Fractals often arise from relatively simple mathematical processes, yet they give rise to incredibly intricate and visually stunning patterns. This juxtaposition of simplicity and

complexity is a hallmark of beauty in mathematics.

3. **Artistic Expression:** Fractals have inspired artists and designers to create mesmerizing visual art that reflects both mathematical precision and aesthetic appeal.

4. **Nature's Design:** Many natural phenomena exhibit fractal patterns, from the branching of trees and veins in leaves to coastlines and mountain ranges. This suggests that fractals are a fundamental aspect of how nature organizes itself.

5. **Chaos and Order:** Fractals embody the balance between chaos and order. Despite their intricate complexity, they arise from simple iterative processes, illustrating the delicate interplay between deterministic rules and emergent patterns.

6. **Mathematics in Nature:** The presence of fractals in nature suggests that certain mathematical concepts aren't just human inventions, but fundamental principles that govern the universe's design.

7. **Multidisciplinary Influence:** Fractals have impacted a wide range of disciplines, from mathematics and art to biology, physics, and computer science. They've led to innovations in computer graphics, data compression, and more.

8. **Aesthetic Appeal:** Fractals appeal to our sense of beauty and wonder, drawing us into their intricate structures and encouraging exploration.

9. **Educational Value:** Fractals provide accessible entry points for exploring mathematical concepts, making complex ideas tangible and engaging for learners of all ages.

In conclusion, the beauty of fractals and self-similarity lies in their ability to bridge the gap between mathematics and the natural world. They inspire awe, encourage exploration, and highlight the deep connections between simplicity and complexity. Whether in the patterns of leaves or the algorithms used in digital art, fractals and self-similarity serve as a reminder that the intricate beauty of our universe can be understood and appreciated through mathematical lenses.

Mandelbrot Set and fractal geometry's influence on art and design

The Mandelbrot Set is one of the most famous and visually captivating fractals, and it has had a profound influence on art and design. Its intricate and self-replicating patterns have inspired artists, designers, and creators across various fields to explore new dimensions of visual expression. Let's delve into the Mandelbrot Set's impact on art and design:

The Mandelbrot Set: The Mandelbrot Set is a complex mathematical set generated by iterating a simple mathematical formula. It's a visual representation of the behavior of complex numbers under iteration, resulting in intricate and infinitely detailed patterns. The set contains both self-similar and infinitely complex structures, making it a prime example of fractal geometry.

Influence on Art and Design:

1. **Digital Art:** The Mandelbrot Set's intricate patterns and self-similarity have captivated digital artists, who use computational tools to create visually stunning and detailed fractal art.
2. **Generative Art:** Artists leverage the iterative nature of fractals to generate art using algorithms, allowing for the exploration of both order and randomness in their creations.
3. **Visualization:** Fractals, including the Mandelbrot Set, have been used to visualize complex mathematical concepts and make them accessible to a wider audience.

4. **New Aesthetics:** The intricate and self-similar patterns of fractals offer a unique aesthetic that contrasts traditional forms of art, often leading to visually mesmerizing and thought-provoking results.

5. **Patterns in Nature:** The Mandelbrot Set's influence extends beyond the digital realm. Artists have drawn inspiration from its patterns to represent natural phenomena such as coastlines, clouds, and organic forms.

6. **Architecture and Design:** The concepts of self-similarity and fractal geometry have inspired architectural designs that mimic natural forms and exhibit complexity at various scales.

7. **Textile and Fashion Design:** Fractal patterns have been incorporated into textiles and fashion designs, offering a modern and intricate look.

8. **Music and Sound:** Fractal geometry has also influenced the creation of music and soundscapes by translating mathematical structures into auditory experiences.

9. **Educational Value:** The Mandelbrot Set's accessibility and visual appeal make it a valuable tool for introducing people to the beauty of mathematics and sparking their interest in exploring its applications.

10. **Interdisciplinary Collaboration:** The intersection of mathematics, art, and design has fostered interdisciplinary collaborations, pushing the boundaries of creativity and innovation.

In summary, the Mandelbrot Set's captivating patterns and fractal geometry's inherent aesthetic qualities have left an indelible mark on art and design. They have opened new avenues for creativity, driven technological advancements in digital art, and expanded the possibilities for visual expression across various mediums. The influence of fractal geometry continues to inspire artists, designers, and creators to explore the harmony between mathematical complexity and artistic beauty.

Diverse landscape of pure mathematics

The landscape of pure mathematics is incredibly diverse, spanning a vast array of concepts, theories, and disciplines. Pure mathematics focuses on abstract mathematical structures, relationships, and theories, often without immediate practical applications. Here's a glimpse into the diverse landscape of pure mathematics:

1. Number Theory: Number theory delves into the properties of integers and their relationships. Topics include prime numbers, divisibility, modular arithmetic, and famous conjectures like the Riemann Hypothesis and the Goldbach Conjecture.

2. Algebra: Algebra studies abstract structures like groups, rings, and fields. It investigates the properties of algebraic systems, including polynomial equations, symmetries, and mathematical structures that extend beyond numbers.

3. Geometry and Topology: Geometry explores shapes, sizes, and spatial relationships, while topology focuses on the properties that remain unchanged under continuous deformations. Topics range from Euclidean geometry to more abstract concepts like manifolds and homotopy theory.

4. Analysis: Analysis examines limits, continuity, and change. It includes real analysis (study of real numbers and their properties), complex analysis (study of complex numbers and functions), and functional analysis (study of vector spaces of functions).

5. Combinatorics and Graph Theory: Combinatorics studies counting, arrangements, and selection of objects. Graph theory focuses on networks of nodes and edges. Both areas have applications in discrete structures, algorithms, and optimization.

6. Logic and Set Theory: Logic studies formal reasoning and inference. Set theory explores the properties of sets, including infinite sets and their sizes. These areas form the foundation of modern mathematics.

7. Abstract Algebraic Structures: Pure mathematics explores various algebraic structures beyond numbers, including vector spaces, algebras, and modules.

8. Category Theory: Category theory abstracts mathematical structures and relationships, providing a common language to describe various mathematical concepts.

9. Representation Theory: Representation theory studies how groups can be represented as linear transformations, finding applications in physics and algebraic geometry.

10. Mathematical Logic: Mathematical logic examines formal systems, the limits of formal proofs, and the nature of mathematical reasoning.

11. Model Theory: Model theory studies the relationships between mathematical structures and their interpretations, with applications in algebra and geometry.

12. Differential Geometry: Differential geometry studies curvature, tangent spaces, and other geometric properties of smooth manifolds, with applications in general relativity and more.

13. Algebraic Geometry: Algebraic geometry studies the solutions of polynomial equations and their geometric properties.

14. Representation Theory: Representation theory studies how groups and algebras can be represented as matrices, leading to applications in physics and algebraic geometry.

15. Homotopy Theory: Homotopy theory studies topological spaces' properties under continuous deformations, with applications in topology and algebra.

This is just a glimpse of the diverse landscape of pure mathematics. Each area has its own beauty, intricacies, and connections to other areas, showcasing the richness of mathematical exploration.

Continuous nature of mathematical exploration and its impact on understanding the universe

Mathematical exploration is a continuous journey of discovery and understanding that has a profound impact on our comprehension of the universe. As mathematicians delve into abstract concepts and relationships, they unveil the underlying structure of reality and provide insights that extend far beyond mere equations. Here's how the continuous nature of mathematical exploration influences our understanding of the universe:

1. Unveiling Hidden Patterns: Mathematics allows us to identify and understand intricate patterns and relationships that might not be immediately apparent in the natural world. These patterns reveal the hidden order and organization underlying complex phenomena.

2. Bridging the Abstract and the Concrete: While mathematics starts with abstract concepts, its applications frequently lead to concrete real-world implications. This process bridges the gap between theoretical exploration and practical understanding, impacting fields like physics, engineering, and economics.

3. Inspiring Scientific Inquiry: Mathematical theories often serve as a driving force behind scientific inquiries. Predictions derived from mathematical models lead to experimental tests, allowing us to validate theories and gain a deeper understanding of natural phenomena.

4. Advancing Technological Innovation: Mathematical principles provide the foundation for technology development. From cryptography to computer algorithms, mathematics drives innovations that shape our modern world.

5. Expanding Frontiers of Knowledge: The continuous exploration of mathematics pushes the boundaries of human knowledge. New discoveries often open doors to uncharted territories and inspire further research and inquiry.

6. Guiding Physical Theories: Mathematics plays a vital role in the formulation of physical theories. Concepts like calculus, differential equations, and tensors are essential tools for describing and predicting the behavior of the universe.

7. Understanding Symmetry and Harmony: Mathematical symmetries found in nature and represented by group theory provide insight into the harmonious structures underlying the universe.

8. Exploring Beyond the Observable: Mathematics enables us to theorize about phenomena that are beyond direct observation, such as the behavior of particles at the subatomic level or the expansion of the universe.

9. Unifying Diverse Concepts: Mathematical concepts have the power to unify seemingly disparate ideas across different disciplines. For instance, concepts like tensors bridge the gap between physics, mathematics, and engineering.

10. A Shared Language: Mathematics serves as a universal language for communication and collaboration among scientists and researchers, enabling cross-disciplinary dialogue and knowledge exchange.

11. Philosopher's Stone of Reality: Mathematics has been called the "philosopher's stone of reality" because it has the unique ability to distill complex phenomena into elegant equations, revealing essential truths about the universe.

In essence, the continuous exploration of mathematics transcends disciplinary boundaries, influences scientific endeavors, and profoundly impacts our understanding of the universe. It provides a pathway to uncovering the mysteries of nature, revealing the inherent order and beauty that underlies the cosmos.

www.ingramcontent.com/pod-product-compliance
Lightning Source LLC
Chambersburg PA
CBHW062247290526
45794CB00006B/2436